Predicting the Weather

Written by Brenda Parkes
and Judy Cooper

Table of Contents

How's the Weather?

The weather is always changing.

One day it can be beautiful and sunny, and the next day it might be pouring rain.

In this book, you'll learn how scientists **predict** the weather and why people depend on them.

If you knew it was going to rain, would you want to go swimming?

Weather Watchers

People like farmers, fishers, and construction workers depend on good weather. Weather reports warn them when bad weather is coming.

If farmers know a storm is on the way, they have time to protect their animals, and fishers will know not to go out to sea.

Weather reports can also warn you if a winter storm is on the way or if it's a good day to go to the beach.

snowplow driver

Weather Patterns

Some weather can be predicted long before it happens. The seasons are predictable **weather patterns**. They happen much the same way year after year. We know that it will be hot in summer and cold in winter. But we don't know how hot or how cold it will be on any given day. That's the meteorologist's job.

Groundhog Day

According to an old myth, a groundhog named Punxsutawney Phil can predict the start of spring. On February 2, if he comes out of his hole and doesn't see his shadow, it's a sign that spring is here.

winter

Does winter mean it's time to put on a warm coat where you live?

spring

Do you wait all winter to see the beautiful flowers of spring?

summer

You can usually count on warm, sunny days in summer.

fall

In your town, do the leaves turn colors in the fall?

Causes of Weather

As the sun shines on Earth, it warms the air. The air in some places gets warmer than it does in others. These large warm and cool **air masses** move from one place to another.

As they move, they may change the weather where you live.

The sun provides heat.

Something in the Air

Changes in weather are caused by heat, moisture, and the movement of air.

summer tornado

Oceans and lakes provide moisture.

Moving air brings temperature changes.

When cold air masses meet warm air masses, it usually causes rain, storms, sleet, hail, or snow. Afterward, the weather clears and temperatures will get colder or warmer over your area.

thunder and lightning

The turning of the earth pushes the air masses in different directions.

If you know what kind of air mass is headed your way, you can make a weather **forecast** for the next few days.

How does the weather affect your plans for the day?

Meet the Meteorologist

A **meteorologist** is a scientist who studies the weather. Meteorologists make weather forecasts that tell what kind of weather to expect. Some of the weather reporters you see on television are meteorologists.

Television weather reports can help people plan their day.

Look at some of the tools meteorologists use to gather information about the weather.

Weather buoys at sea send wind speeds, temperature readings, and other information to land.

Satellites in space send pictures of weather patterns, like this hurricane, back to Earth.

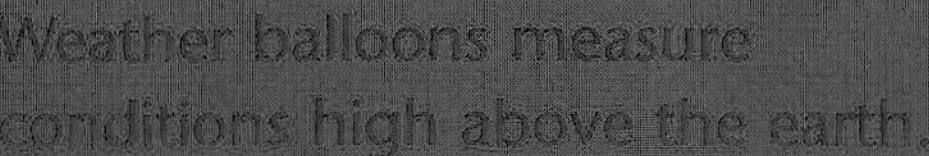

Weather balloons measure conditions high above the earth.

Radar trucks can get close to storms and scan them for information.

Reading a Weather Map

Reporting the weather can be tricky—it changes all the time. Weather maps help meteorologists keep track of what's going on across the country.

Look at the weather maps. The **map key** tells you what all the symbols mean. What's the weather like where you live? What do you predict it will be like tomorrow?

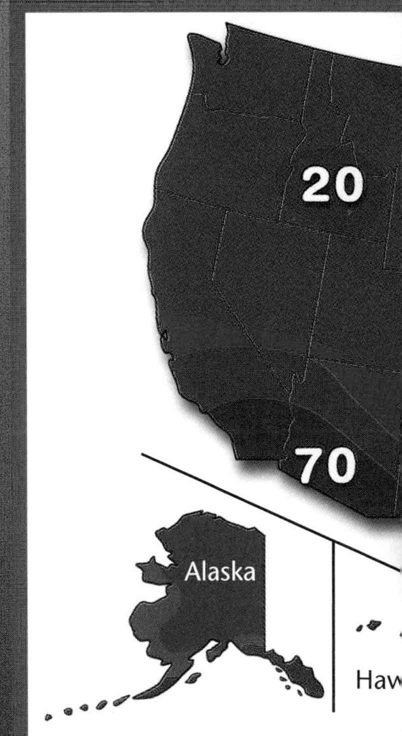

Map Key

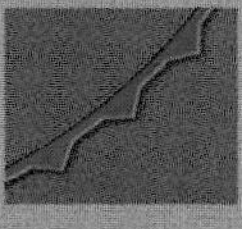
= cold front; brings cooler air (arrows show which way it's moving)

= warm front; brings warmer air (bumps show which way it's moving)

= high pressure; brings dry, sunny weather

= low pressure; brings clouds and rain, snow, or storms

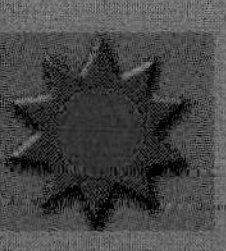
= sunny

= partly cloudy

= cloudy

= rainy

= snow

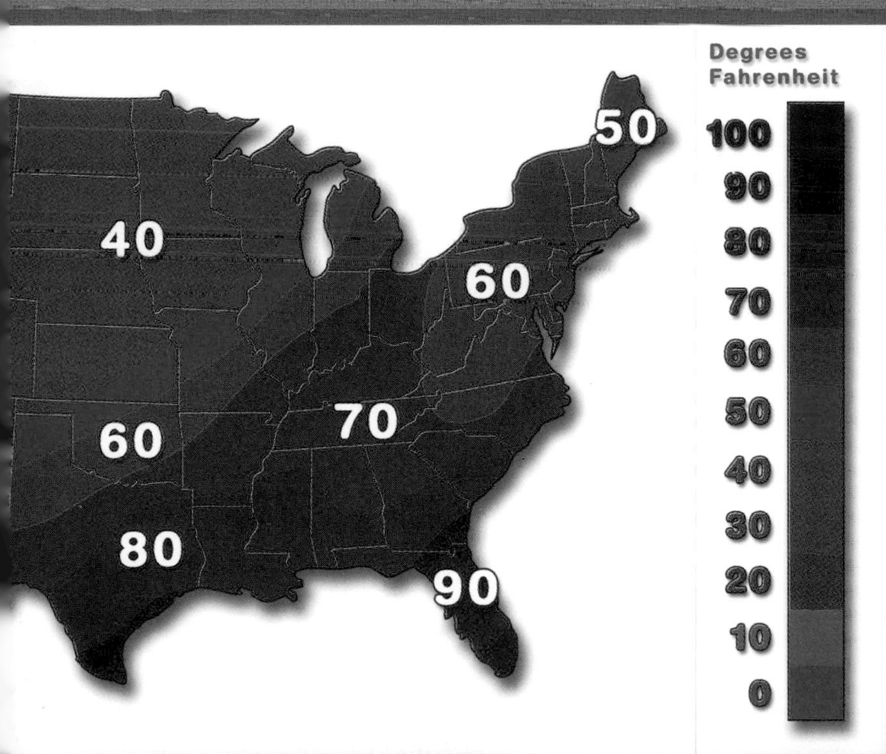

Watch the weather report on television tonight. See how many of the map symbols you already know.